www.cagliastrotheironring.com

www.northseatales.com

Advanced Class Transcript Books, a new series by North Sea Tales Inc. featuring transcripts of ADVANCED CLASSES in the techniques of the Science of Sorcery.

THE ALL WHITE

Two Expanded Advanced Class transcripts with Instillations discussing the location known as THE ALL WHITE, Thoughts, Ideas, Intentions, Mimicry, Self-Effigy, Require and Desire, and Timeline Manipulation

By SORCERESS CAGLIASTRO

This is a set of class transcripts. It is slightly edited and profoundly enhanced for clarity, and any names of students is removed. These transcripts are in the language and speaking style of The Sorceress. This version is enhanced with additional material and comments by Sorceress Cagliastro.

THE SIGIL FOR "KNOWING"

This book is a product of
NORTH SEA TALES PUBLISHING INC.
www.northseatales.com

THE ALL WHITE, Science of Sorcery and any images or sigils herein are protected by copyrights 2010, 2011, 2012, 2013, 2015, 2016, 2017, 2018

COMMENTS FROM THE SORCERESS CAGLIASTRO

There are two extremely enhanced transcripts in this book. The first one, THE ALL WHITE, Thoughts, Ideas, Intentions, Mimicry, Self-Effigy, Require and Desire, and Timeline Manipulation **addresses a location in which Sorcery is worked.** This place has been taught to my PRECIOUS Advanced Students, and has not been written about prior to this publication. Some of the labels and language will overlap from one transcript to the other – however that is the nature of the work. In the Science of Sorcery, as in life, overlaps and accumulations are the stuff of increased knowledge....

It is my intention to write further material on THE ALL WHITE and other subjects mentioned herein, however until that time, this is available....

THE ALL WHITE

Two Expanded Advanced Class transcripts with Instillations discussing the location known as THE ALL WHITE, Thoughts, Ideas, Intentions, Mimicry, Self-Effigy, Require and Desire, and Timeline Manipulation

By SORCERESS CAGLIASTRO

THE ALL WHITE

WNCC – 2-7-17 – In the White Room...
So it is the 7th of February, 2017 and this is the Wednesday Night Core Class. I want to talk about what we're doing here and the **ALL WHITE**, and all of these sorts of sensations and experiences that have labels, which are new....

We have talked about BLACKHOUSES and we have talked about the BOXES and all kinds of DIRECTIONAL SORCERY and this sort of thing, but lately the work has evolved into a place where we're using a more tactile environment, the ALL WHITE being this location. If you would like to call it a room, you can call it a room; I prefer to call it the ALL WHITE... this idea that we have from at least my point of view; my end of the thing the Gold Lion - which is what my Sorcery looks like, (covered in an upcoming separate book) and the furthering of that idea which borders on an **intention** - and that's kind of what I want to talk about in the beginning of this thing tonight; this **idea** that your Sorcery **looks like something**.

That it, YOUR Sorcery, becomes so powerful that it **becomes** so fully enriched with process that it **becomes** a thing. This may be best accomplished by utilizing Mimicry and Self-effigy.

Some of you have talked to me about the sensation where you're not sure if what you're seeing, or what you're feeling or doing is real. And by **real** I mean - if I'm going to define real - what I'm going to say is I have a **thought** or an **idea**, and I enact it in some way and **then** there is an outcome, and that can present itself in the immediacy or the longer-term. So, for example, when we're working with Daemons and I say that there is a Daemon in the room, do you see it, or does your **mind provide for you**, a visual experience that **allows** you to play along - and I don't mean that in the light weight at all, but to play along in the process. This can happen and in the work of teaching the techniques, I must find within myself to show you the difference between what you see and hear – and what your mind provides for you to ALLOW yourself to THINK you are seeing and hearing actual entities.

> This is vital because this is **NUANCE** - this is the tiny **space** between idea and intention.

This class group, as you know, is beyond precious to me, and I teach things in this class that I never even thought I would say out loud and share, and I teach these materials with a joy and a requirement to continue doing that. So I will expand here on one of the real anchors of the work, an anchor – not a Pillar. **The categories of Thoughts, Ideas, and Intentions are very important and quite interesting because although they are technically categories, like all things in Sorcery, they have a relationship through almost a tunnel, or a passageway from one thing to another.** And yes we have talked about Thoughts Ideas and Intentions quite a bit, but I have a desire to go back and touch on this because it is vital that advanced students – or even those who are deeply aware of my work, understand in the most practical way.

Let's suspend a desire for details and just work through this opportunity.

DECONSTRUCTING THOUGHTS, IDEAS and INTENTIONS

So I have a thought that is - **I feel like traveling**. I have this thought… and it's a thought because what it is, is sort of an unformed thing, but it **IS** attached to desire in some way. It's not attached to requirement yet, but it's attached to desire. This thought is… **So I feel like traveling**. I have a… I don't know, sort of a craving to go somewhere. It's a thought. And then I think, well I could go to a local shop, or I can go to Scotland, or I can go to Italy, or I could go to New York, because the thought is involved only in **the process of moving from one place to another in an experience called; going somewhere.** And then you get into the idea… okay so I picked New York for example… And I say; I had a thought and I decided to go somewhere, and now I have an idea that that place should do some things for me. It should entertain me, it should provide 1,000 choices for where to go for a great martini. It should provide places to go and buy fantastic platform sneakers. I should be able to see my friend, Sebastian.

So now we don't have a Thought anymore, we have an **Idea; a structured experience that we're thinking about having**, and it seems to have **requirements**, and this is how we **move** from desire to require; **a thought being a desire, and when we get past thought, and we get into idea, we get into require**. And when we are in require, we are in the **construction** business. In require, we start to find the pieces that come together, that make what we desired and now required tangible in some way. Now think about a tiny **Nuance**, that sliver right there... I'm not in New York in that moment, but I'm **putting together these constructions, the sort of scaffolding**, the work **around** the story. I'm looking at a train ticket, or I'm booking a hotel room, or I'm deciding; well it's a little bit warmer in there, I don't need this big coat and **I'm doing all these things that your timeline does not know didn't happen yet**... because your timeline is this tickertape that's running underneath you in your life; back and forth and as Sorcerers we get to play back and forth, and that's the difference.

So if we stay in this tiny **Nuance** which is where I desire to stay for a minute, where we've had a thought, we're in an idea and we're flipping over into intention; **right in that corridor between idea and intention,** and what does that look like? All right, so I've booked a train ticket and maybe even booked a car. **So I decide that I'm leaving Friday morning and I'm going to New York, and now any period of time between right now and getting on the train is different**. It's a transition period in a sense, and hold that for a minute and think about it... between idea and intention, because "thought" we're done with... thought played the game, it brought us to this story... So now I'm in this transition between idea and intention, **but am I?** Yes, Because I **can't** be in a place now where I **didn't** think I was going to New York for the day, or the month, or the rest of my life... or whatever... however that works out. That place of non "going" no longer exists. There is only the place of going – and a sensation of "looking forward" to something....

So we don't really have absolute stops. We can't say; I'm on the corner of this and this, but I'm not heading towards the corner of this and this, because as we move from thought and idea into intention, we are actually peddling in the next fact, if you will. **Once I buy a train ticket I'm in New York**, because my timeline has now presented itself in the process of manipulating everything between now and then.

> **AND THIS IS HOW YOU CAN CONSTRUCT THE ABILITY TO LEAVE THINGS BEHIND – THE ART OF WALKING AWAY.**

Now, when you move into intention you get in the car and you go to the train and you step foot on the train. Now you decide, as a Sorcerer if booking that car or getting in it, or getting out of it, and getting on the train; which one of those things makes the next process; intention, unstoppable?

Well, if the train is going from wherever you are to wherever you are going **nonstop,** then the concept is; aside from you pulling some sort of medical emergency, which you could do; it's unstoppable.

That is the visual that I'm setting up, and you know I don't like use this idea of visualization, because I don't want to, you know; "talk show" lingo it up, but that's the sensation of **placing oneself in the unstoppable**.

When we sit and we say; come into the ALL WHITE, or come into the BLACKHOUSES, or I'm going to create a room for you to step into or you, or you do this yourself, you are already in the process of **allowing** for whatever it is that you built that room to do, to come in. So if I say; I'm building this room and in this room is one of my attending Daemons for example 7; none of you really know that much about what 7 may look like, and you get in the room, and you know **that 7 is coming in and is going to work on something with you, then you're already in the process of being with that Daemon**, because there's a collective consciousness in this that is very powerful once an intention of it is already under way.

So here is a set of sigils we will use tonight. I refer to these as a set of sigils, because in this case I don't feel they are as functional when utilized

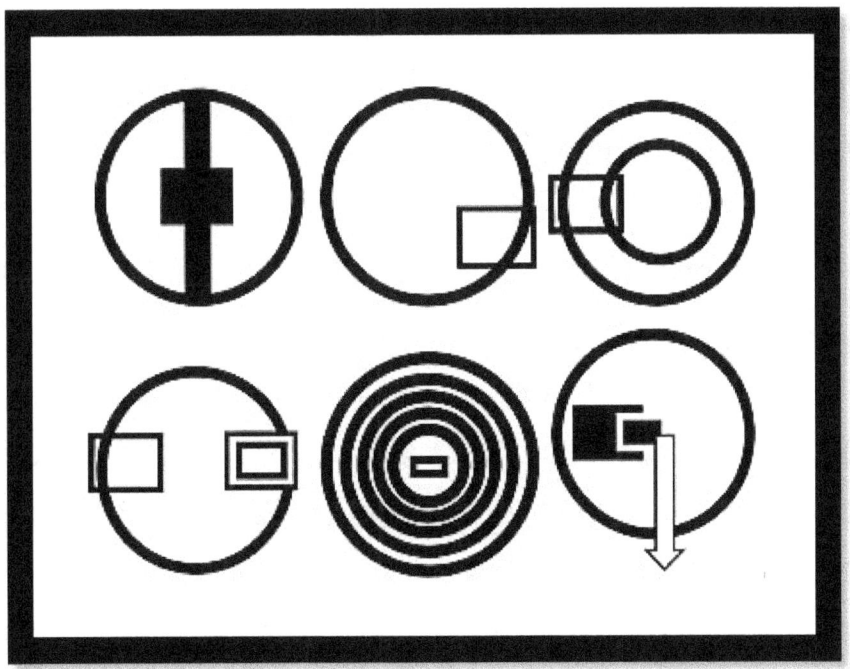

separately. Later in this class and in the next one (also in this book) I will say to you tonight, that when we start this process; when you look at the first sigil that when there are no boundaries you are already in the last one as well. When you are on the third one you're on the second one etc. because the building of this set of sigils has to do with a solid commitment to knowing limitlessness.

Sometimes in sorcery, as you get; I don't know that the word **better** is the right choice, but perhaps as you develop more skills or become more comfortable with your own process, what happens is; you forget all of the processes, all of the things that feel like that ugly phrase that I really hate which is **'the steps'**. Why do I hate that phrase so much? Because if you can actually tell me how small you can break something down into "steps", I will listen to you. I have yet to hear how that process works. If I stand up and I walk across the room, you can say; well you stood up and you started walking, or you could say; you sent a message to your muscular structure where you leaned forward a little bit and then you were kind of standing up, and you kind of straighten your knees.... So where is the break off? Where is the tiny-ness? How small can it be – and therefore where is the smallest Nuance? **I ask because that is where we work – in an understanding of the nuance of the moment....** How small can it get? Because as Sorcerers we have to ask ourselves how small can it get?... What is the smallest "step" in walking across the room - is the answer standing up from sitting down, is it leaning forward, is it putting weight on a foot? Yes, I am using the language in a sort of on uncomfortable way, beginning sitting down and then standing up; between those action there are many processes. One does not sit down and then finds oneself suddenly standing, at least not at the anatomical level.

> **The explanation of how we cannot logically find the smallest nuance is this. In process, a thing happens. There is always a thing… and that 'always thing' is the energy between thought, idea, and intention. It is the energy between; something is happening in a game somewhere and you want to change the outcome so you get to the end of the game before it starts - and you are in the celebration of the winner's circle, and then all of it fills back in.**

I saw today on some news channel, I think it was Fox; I think her name is Kennedy; she said something very funny. I taped it... well I recorded it on my phone because that's how I now do things, and what she said was; (I am paraphrasing and leaving names out)"something happened at the Super Bowl, so-and-so did something and then they won, and personally I just think it was all sorcery." And I thought; well you know, you might have a point on that. I know a lot of people who worked on that game. So at which point in a game like that, and understand that I know less about football than almost any living human being, but what I know is, and I'm using this as an example, because it's the thing we've all know happens once a year, is that those who desire that win - what they do is **get** in the winners circle (and coming from a family that had some jockeys in it, I only think of winners circle,) because getting in the winner's circle with the winning horse, and picking up the big prize **BEFORE** the game starts **is the only way to win. That is how Sorcery is placed into action – and that is what makes assumed boundaries regarding timelines irrelevant to Sorcerers.**

Breathe.....So where does the process break itself up? Does it break itself up? I'm not sure that it does at all.... If you can explain to me where the tiny second is that stands alone I will explain to you how that second breaks down into a thousand or a million other moments...even more than my human mind can see.

Consider this; consider sexual gratification. Consider the moment of orgasm. And I talk about this a lot because it's something we all have our own personal connection to and no two people experience anything one hundred percent alike, so this is a very personal thing. When one feels orgasm coming on, one is **already** in it, and when one is experiencing it fully – then there are a zillion uncountable moments referred to as a time period. So where does it cut off? Where is the moment where that pleasure isn't happening in that sort of arch of beginning to end? **One cannot sliver out a moment tiny enough to define it**... remember that. One cannot sliver out a moment tiny enough to define it. We have language that defines time however it is for convenience.

It is not meant to be accurate – but rather to convey an understanding of pacing. We say things like; *I'll be there in a minute. Give me five seconds. I'm standing up. I am almost there. This should be over soon. Things are happening.* We have language that allows us to corral events into understandable pieces of action that we can define, but what we don't have is a definition of **absolute sliver of nuance.** The fact is… and as advanced students you know that I love the word fact because it has profound qualifiers, which are rare so therefor I rarely use it… The fact is that we don't have that absolute sliver, and therefore as Sorcerers… and hear this…

> **Every single thing that we do in Sorcery will absolutely connect to what's next. And that is all we know for sure….**

Think about voids; what falls into the void? My student's, and anyone who has read my work, knows this:

That which falls into the void, without intentional manipulation is ALWAYS that which is closest, most familiar, and heaviest. This is the reason we relapse, the reason we repeat, and the reason we cannot break toxic cycles.

If what happens in an orgasm is; you start to feel pleasure and then your body is allowed to do whatever it does independently - it's just personal to everyone… it knows what to do next – then orgasm is achieved. If when you start to feel an orgasm building – and you fall back into the realm of guilt or shame or confusion or whatever crap you are carrying – then the void opens up – swallows the pleasure – and you do not climax. This is an illustration of what happens when we do not utilize INTENTIONAL MANIPULATION = SORCERY – to make sure what falls back into the void is NOT whatever is closest and most familiar, but rather whatever is DESIRED by the individual in the moment.

Back in the example of -If you are going to New York you get in the car, out of the car, onto the train, sit down and face in a direction that is pleasing to you to move towards. So what's next? **We are Sorcerers, so we are in the business of what's next.** The **Four Pillars** and all of the techniques push us to what's next. **Directional Sorcery** bonds, and breaks, and pushes, and shoves, and structures appropriate timing, and takes out time, and puts in time, and closes in, and pulls away because it's all about what's next. If you think about **everything** that's next as **a void that's popping open and presenting an opportunity**... an opportunity in real time, not something to be concerned about, but **instead of "the usual" as an opportunity to manipulate and therefore choose whatever is closest, heaviest, and most familiar**, then that's what your Sorcery **is**, and what it delivers by **CHOOSING** what's surrounding that void – by **CHOOSING** what fills it, by **CHOOSING** what is **NOW** the **new thing/s** that will become familiar and fall into the void - **and you break patterns and get to move forward**. Even if it's fixing the Super Bowl, fixing an election, doing some healing... it doesn't matter... getting a car that you want, getting somebody to buy you a drink in a bar when the person wasn't even looking at you; whatever it is, that's how you do it.

> **Use the BATTERY of the failed non-satisfactions to give you the roadmap to CHOOSE the new content.**

– and that's how you get it.... how you get everything you desire.....

Let's talk about the ALL WHITE - the "room" for a minute. The ALL WHITE.... What I'd like you to think about is **a place that does not have any boundaries**, IN FACT (there is that word again making a second rare appearance) a place that is not even **CAPABLE of sustaining boundaries**. <u>How could you be a place that does not have any boundaries and see it?</u> Think about that, how can you see it? If I said to you; I have set up a room that was 10' x 10' and it was ALL WHITE painted inside and there was nothing but white, white, white, white, you can deal with that because you can stand on a white floor, and you know that the walls are a comfortable space away and you could see it. However, what if all I told you that I was presenting an ALL WHITE? How do you find that? How do you stand in it? Is the air white? How do you find the edges knowing there are no boundaries? The issue is a deeper one than the visual:

> **IF I TELL YOU THE ALL WHITE IS A PLACE WITH NO CAPABILITIES OF EVEN HAVING OR UNDERSTANDING BOUNDARIES – THEN YOU ARE FORCED TO CONFRONT YOUR OWN CURRENT AGREEMENT REGARDING YOUR LIMITATIONS.**

Do you have boundaries - I don't know... But what you can do...

> ...what you absolutely must do to practice
> Sorcery - is train yourself to
> **NOT REQUIRE** edges; to not require yourself
> to know where one thing stops
> and another one starts.
> You must not pre-set your limitation.

Train yourself as Sorcerers to do this because the second you are able to act without boundaries; to not have that limitation pre-set, then your Sorcery will absolutely move on to your next requirement... and requirement is idea, and intention is fulfillment.

So now that I've gone on and on about restructuring voids and deconstructing boundaries, and sort of clarified this sensation that all of what we are doing is connected to what we **choose** to be "next", and I must clarify. I am not discussing "next" in the context of what is **expected** of us in any social or cultural way. I don't mean, you know; I should want any specific expectation.

And I intentionally choose to crush the mythology of coincidence, karma or the idea that something is 'meant to be' or any other construct that chooses our farthest reach. What I do mean is quite simply that when one begins to enact Sorcery **truly and with profound intention of BEING the winners circle, (BEING – not just being IN),** then thoughts, Ideas, and intention, **when one enacts Sorcery with profound intention;** one has ALREADY PERFORMED THE SORCERY. Some of you have sat with me and just seen me move a finger, and then an outcome happens later at a meeting when that outcome was not likely, or someone is having an issue and I'll simply make a gesture and then they'll call and say the issue is solved. And then some things happen that don't work. Sometimes things are very far into where they are and the gesture to pull them back is more complicated, because maybe the thought, idea, and intention of that process is more profoundly lined up with completion than any way that I know that moment to change it.

> **For this we go to the ALL WHITE and change and confuse the boundaries....**

So what I'm going to do is, I'm going to present the same sigils that I presented earlier. (THE CLASS SAW THESE ONCE IN A PREVIOUS CLASS) I'm going to ask you to remain in an ALL WHITE. And I defined an ALL WHITE in the last class or the one before that however I feel I have done a good job here of relating the environment.

> **I use ALL WHITE because this is a setting that urges your eyes to seek color.**
> **Seeking is dis-comfort, discomfort is movement, movement is energy.**

In the ALL WHITE You find your eyes not sure if they should close. It is destabilizing **and when we are destabilized we make choices.** We have talked about MASQUERADING in the previous class, however I will talk about it via the sigil tonight so you will have enough material to grasp the value. We talked about doing whatever it takes to be in the sensation of ALL WHITE. In the ALL WHITE we **must** be in a MASQUERADE, changing, seeking understandable or familiar function, and we must fight back against the predictability of what happens to the void without our intentional manipulation. This is a perfect place for Sorcerers because it is destabilizing. We are not built to stand in absolute nothing. We are not built to stand comfortably in a borderless place such as the ALL WHITE. We are not built to stand in a place with **little to no stimulation**. We are just not built in that way.

So in sorcery when we go into a place like this, we are truly suspending the **drama** of operating from within the human form. We are talking to the cones in our eyes that are seeking color, and definition, and shape, and we are as saying to them; sorry no fulfillment… but what we are really saying to them as people who practice Sorcery, is that; (and I don't mean this in an esoteric way, I mean this in a way that we answer our biology which is necessary to do when you practice sorcery… we say to them;) everything is already in here. **It's all here because there is no way to pack all of it into one thought, so just except that it's all present.** It's something that one can accept that it is all present because what is happening is that all is possible when one looks at every interaction in every slice of every nuance of Sorcery as a void filled by void, filled by void, filled by void….. Thus the requirement for manipulation of the surrounding materials….

BREATHE……Stretch – have an adult beverage…..

> **When you are dealing with voids what is often closest, heaviest, and most familiar... is another void. This is the physical description of how in sorcery we move from event to event to event to completion.**

If you desire for the team to win then you go stand in the winner's circle before the game, and because it is unlocked regarding the timeline via Sorcery, you become the winner's circle. This is time unlocked Sorcery. But what happened? How did you get there? It's 6 o'clock, this event is over 10... how'd you get there? Well, you got there because **'there'** is already happening in somebody's timeline somewhere whether you have a working understanding or example of that or not.

Somebody has an intention about it, you've connected to it, and all that's happening between now and then is a series of dominoes... of voids falling one after the next, hitting each other, demanding conclusion, demanding satisfaction; pushing, pushing, pushing, pushing, landing, and finally paying off and being resolved.

Sorcery feels like, often, a sensation of a great sort of moment of "YIPPEEYIYAY! I'm so proud of myself. I did this thing, how cool is this wow!". But really all it is, is in the fact; a word I don't use often but I am using it a lot tonight in the subject... it's in fact the sensation of **desire proved in resolution**. "Well, I'm glad that's over and the payoff was magnificent." That's what the energy of Sorcery is saying; glad that's over, and let's get on to what's next. Giddy up!

So what's next for us now is this group of sigils. When we look at them I will ask you to be in the ALL WHITE and to allow yourself to fully develop and define that, because it's a working space. Sorcerers who do advanced work require a solid "build" of the ALL WHITE to fully engage.

Previously as a class we've talked about the BLACKHOUSES however they are a much more constructed set of locations with predesigned functions, and we've worked only once in the ALL WHITE. Going forward as a class we will be working for at least a few more times in the ALL WHITE as we move through the process of understanding the utilization available in it. I sent these sigils to you before class and I suggested that you look at them, and then I did say that ultimately you will telescope them all and allow for them to present some sort of composite. Well I ask this group, and this group only, and not anyone else who reads my books or has worked with me briefly - to get over yourself about telescoping. You are advanced students and by now you should know that **ALLOW** is the big ticket.

Telescoping is important because it allows your mind to "sift" through the aspects of the sigils, (or anything you are refining to your desires) resulting in the remaining portions which are directly associated with individual personal requirements.

The thing about telescoping is that it happens whether you like it or not, or intend it or not. The effort to learn to do it intentionally it is stressful and complex and torturous and bullshit. What happens is something like this. First you see your friends face and then you see the friend has glasses on, then you see someone else with tinted glasses, and then when you look at your friend and you see that the friends glasses appear tinted - and you've telescoped. Or you walk down the street and you glance at the glass in a shop window – **at** the glass not **through** the glass, then you glance away and see someone you are attracted to. Next you look back at the glass and see through it and notice a hat in the window. Then you look back at the attractive person and for a moment it appears they are wearing the hat. There - you've telescoped. You've taken something beautiful and added to it to **customize** it in some way. Another type of telescoping occurs when you're walking toward someone you love or someone you find enticing and you start to **not** see anything between yourself and that individual. You're telescoping... you don't think you are, you think you're not telescoping but you are telescoping. What you're doing is placing mimicked layers of that person closer, and closer, and closer, to you and over-riding the space. So, whether it's discomfort, confusion, enticement; it doesn't matter. What we're doing is we're telescoping. We're doing it all the time.

So I will go through these sigils and tell you their stories.

The Sigil called GLUE

The first one being GLUE and in this story; the sigil story, square and round are really the whole story. And we're looking at a thick line, and now you can say that thick line is either crossed over by a sort of rectangular box, or you can say it's held together, or you can say it's two lines held together by that box which is how I perceive it, thus GLUE.

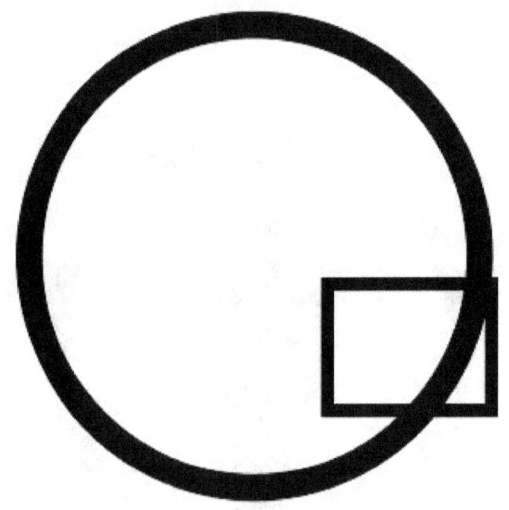

The Sigil called PARANOIA

The second one being PARANOIA and in this sigil what's happening is questions are offered; is that box escaping? Is the circle escaping? Are the square and the circle attempting to combine? Are you the circle, or is it coming in? Is it is getting out of your way, or is it coming to get you? Or is it behind, peering in? And now you're all seeing it that way. You're all seeing it as behind, peering in. What is the story of that square, of that rectangle? What's it doing? Is it entering? Is it leaving? Is it not allowing for peace? Because that's really what it's doing. It is interrupting what looks like a very simple gesture. The circle, the void, the Sorcerer; it's interrupting. You don't know if it's coming in or out because paranoia exists not because someone is following you, but because you know they can.

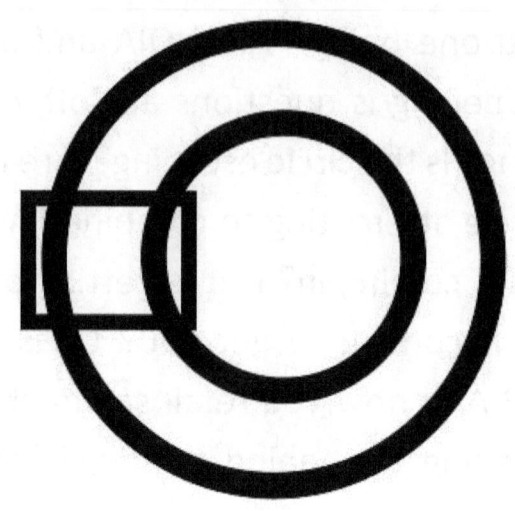

The Sigil called MASQUERADE

MASQUERADE is a very interesting portion of this process because it's entirely artificial, and yet absolutely necessary. If we go out in the rain and it's drizzling, we put up an umbrella... why we do that? If we don't put up an umbrella will we die? Will we have acid burns on our heads from drizzle? No, we will just be damp, but we do it anyway because we have an agreement with comfort.

MASQUERADE simply supports and enhances your agreement with comfort. Now some of us have a greater agreement with comfort, and some of us have a lesser agreement with comfort, so let me talk about that for a second. I had a client who said to me; "oh, you know I'm an outdoors person. I'm outdoors all the time. I don't need all the trappings and the big house and…" I interrupted and said really? How expensive is your camping equipment? He said; I have the best camping equipment a man can buy. So, I said okay, so there's your masquerade, and there's your attachment to comfort. I wasn't judging him, I was responding to the METHOD he used to convey his position. He used the words "I don't need all the trappings and the big house". I challenged him NOT because I took issue with the cost of his equipment, but rather because he positioned his relationship with comfort in such a way that put the weight on his COMPARISON of comfort to wealth. I am a fan of quality and beautiful things and took no issue with the cost of his equipment. I was evaluating only the sensation of his determination to seem unattached to wealth. I was deconstructing his Masquerade.

How are we Masquerading? And because we're talking about Sorcery and not the nonsense of social convention, **there is nothing wrong with a good Masquerade. There is a useful place for it in Sorcery. It plays into MIMICRY**. Masquerade only works contrary to us if we buy into our own story. (believe our own bullshit) In other words, we can take part in the Winner's Circle when WE own the Masquerade, and not when the Masquerade owns us.

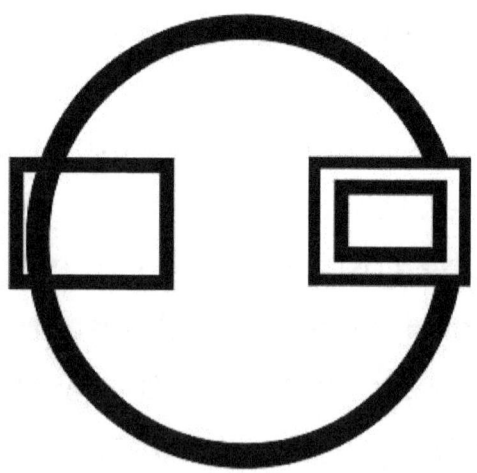

The Sigil called COMPARISON

COMPARISON can be a deadly game if you are not stacking it in your favor. Like all aspects of Sorcery – you have to own the game and the beast that plays it. Comparison is a two edged blade with a resting place between them. If you compare your Sorcery to my ability to enact Sorcery, you might get in a place where you say; why is it not working for me quite the way it works for Sorceress C? You are placing yourself in the lane of lesser experience and therefore disappointment. However if you compare your Sorcery to a guy who might have just started studying and is new to this, then you could say; well you know I'm a bad ass, and I know what I'm doing. The center of the blade, the resting place is of course also available. You can look inside yourself and you can compare your own ability to progress from where you were to where you are now – and ultimately to the pleasure of being able to actualize a bigger future through the use of Sorcery. In the ALL WHITE there are no borders or boundaries. This is the perfect location to do this work.

There are two illustrations inside of COMPARISON. The plain one, the one on the left is comparing itself, its progress and its place (where it is in its life) to another illustration which is more ornate. The more ornate one may look like one representing more work – however think back to the outdoors person with the comparison issues. The center circle represents your "timeline" or how you perceive how much work you've done and how that fits into accomplishment. This is the third method – COMPARISON with the self.

The Sigil called WONDER

Now let us look at the sigil for WONDER. The thing about WONDER, (and one of you asked me if there was a sigil coming for WONDER, so here it is), is as much as these sigils all require each other to learn how to use them, some of them have a better ability to stand alone. WONDER has a better ability to stand alone than the other sigils in this set. I said; in previous instruction that Sorcerer's must **WONDER hard.** WONDER in a limitless way like a child. WONDER in a way that allows you to blast through the doctrine (poison) that you carry. WONDER if you can fly. WONDER if you can produce the job, the house, the lover, even further... the self that you desire and require. WONDER like that... don't bother to wonder if you'll make your train. That is easily handled with Sorcery. WONDER much bigger, harder, stronger... WONDER. And what happens is, if you look at WONDER, **the void is the rectangle; it's very deeply hidden...** and it's very deeply hidden by all the nonsense that we are told that tells us that Sorcery and Daemons and being a wildly successful deviant doesn't exist. Sorcery, Daemons, and being whatever you can create- these are things that doctrine holds up against us, and hits us over the head with until we are defeated… Comparing that to the sigil of WONDER – it is an effort to destroy us so we can't look for the third, fourth, and fifth circle and get inside and find wonder again. However – we can through this process.

> **I am here to UNLOCK WONDER and give it back to you.**

> **Where better than the ALL WHITE – with no borders or boundaries to UNLOCK WONDER???**

So look at the WONDER sigil... You can have two experiences. You can say; wonder is so locked down that I can't get to it, that I have to jump over circle, and circle, and circle, and as these are not a labyrinth, these are locked inside of each other. Or you can say**; my ability to wonder is that protected, and I will see it that way – reframed in the way one sees cherished things. For the purpose of successful Sorcery I suggest you bend towards the second option.**

The Sigil called BUNDLE AND PUSH

And the last sigil in the group is BUNDLE AND PUSH. What do we have here? If you look to the left we have what looks like something that was perhaps the whole rectangle and interrupted. If you look at the small rectangle in the middle it looks like it's free-floating. **But guess what?? - None of those things are happening. The larger rectangle is in fact interrupted by the white frame; the ALL WHITE of the other rectangle, which on the right, you cannot see because it doesn't have a boundary.** That is where you need to stand; right at that right angle in the white frame. That is where you must stand; **inside the white frame that you cannot see the continuance of because it does not provide a boundary, and on top of the ability to remove whatever is in the way.**

This is the sensation of the ALL WHITE.

That is going to be all I'm going to say about these sigils, and now we're going to go back to the beginning and be in the ALL WHITE, and you will be timed in allowing them to... let's just say have an insertion into your skillset. What I am saying is "let's just say allow the insertion", I would like you to just be in the **ALL WHITE** with them, and know that when we are done we are getting to the end of a series of sigils, a series of sigils that we will use moving forward in a very profound way. At the end of this you will be suggested to telescope.... Well, was that the most nonsense you have ever heard in your life? When I say **suggested to telescope** all I'm asking you to do is **allow**. I'm only asking you to take in these sigils, to be in a space with them that is **without boundaries** (ALL WHITE) like the white frame on the final one in BUNDLE AND PUSH, and to learn to interact with them in a way where they bring out the most satisfying facets of this portion of the work in you.

There's a question I'd like you to ask yourself during this process. I'm going to send into the space one of my Attending Daemons, and the Attending Daemon I'm going to send in is Sheeg. Many of you have seen Sheeg. For those of you who haven't, I will tell you exactly what she looks like; she looks very similar to the girl that climbed out of the TV in the movie, The Ring. Or who climbs out of the well. She is small, she is thin, she has what appears to be long stringy black hair, she is in a white mess of a frock that for some reason, (and I don't know why) is often wet. Sometimes you see her feet, sometimes you don't. Her skin is very, very white and sometimes it has some sort of hue that there may be something beneath the surface. She is sometimes faceless, and sometimes in the presence of a rather horrified tortured face, and she is in a state of constant anxiety and chaos. She does not speak words; she screams. I, for reasons of which I do not have an explanation, I am able to interpret those screams as language. She is unfriendly looking, she's cold and damp, and when she touches you, you know you've been touched by something horrific. Sometimes she scratches. She, on very, very, very rare occasions bites. She is very "about bone"... she is very interested in the touching of bone, and she's very interested in the touching of teeth. Sheeg is an ugly scary damp cold creature. She's all Daemon; she's everything we think of when we are five years old, alone in the dark, and we're scared, and yet she will bring to you whatever you require, if you develop a relationship with her such as I have. She is specifically **MY** Attending Daemon. She enjoys very much participating in this work, and she is going to be in the ALL WHITE with you tonight. Now the question I'm asking you to ask yourself... and it is only one simple question; during this work tonight as yourself do you see her, or did you develop her through my description of her? And the way we will know the difference is this - I have placed around her neck a necklace, and it is something that I will not provide a description of until I get emails and find out what you saw, but I will tell you that it is visible, that you can see it when you look at her, and that it isn't a hidden thing. So, when you ask yourself; Do I see her? Take care not to construct the necklace. I'm giving you the opportunity, if you will, to find the place between 'I think I see her', 'I've decided I see her', and 'I see her'.

Okay I'm going to time you, we are going to spend two minutes with each of these sigils. Allow yourself to be in the ALL WHITE. It is not an absolute requirement, but I would suggest that you consider taking a pulse to see if you have had an elevated Blood activity during this process. I will set that up for you now... in the way that we do. you have 10 seconds to find your pulse... to find a comfortable read of your pulse, and begin counting now... (readers set a times for 10 seconds on your phones to accomplish this process).

Stop. Write down your pulse if you chose to take it... and also write down whether it was constant, erratic, hard-to-find, faster than usual; whatever notes assist you in knowing what it feels like to listen to a description of what Sheeg looks like, because that's what you just did.

Okay the ALL WHITE is set. **Allow** yourself into it and **allow** yourself to be in the presence of the **ALL WHITE**, and Sheeg, and the first sigil which is GLUE. **(readers set a two minute timer on your phone to get through this insertion/exercise)** Allow for the combination of these elements for two minutes in the ALL WHITE starting now. Remember to ask yourself the question, to look for Sheeg's necklace, and also you might notice that time feels like, (does it **feel** slower or faster during certain sigils when we're still counting the time).

 (2 minutes – STOP)

There is no break between sigils.

(READERS START YOUR TIMERS)

We are in the next sigil PARANOIA. Sheeg is not above interacting with the sigil itself. I've been asked to mention that. She has mentioned to me that sound was her nails, she feels the floor... that's the message.

(2 minutes – STOP)

(READERS START YOUR TIMERS)

We're in MASQUERADE, so remember to continue to ask yourself the question and take notice if you have any other experiences tactile experiences cold, hot, the sensation of a particular scent, etc. Continue to look for Sheeg's necklace.

(2 minutes – STOP)

(READERS START YOUR TIMERS)

We are entering COMPARISON. COMPARISON is a sigil within itself comparing two sides of the two options. Remember to ask yourself the question. Notice when the time feels longer or shorter for each of these two minute segments. 2 minutes in COMPARISON.

(2 minutes – STOP)

(READERS START YOUR TIMERS)

WONDER, which as you know also has two frames of reference. Is you wonder trapped or is it protected, and while you are exploring that allowing it in, remember to ask yourself the question. Allow this visual sensation of the sigil to have its own effect if that helps to consider it.

(2 minutes – STOP)

(READERS START YOUR TIMERS)

We are in BUNDLE AND PUSH. Stay in the ALL WHITE.

Now I will ask you to just look at the box with all six sigils present inside.

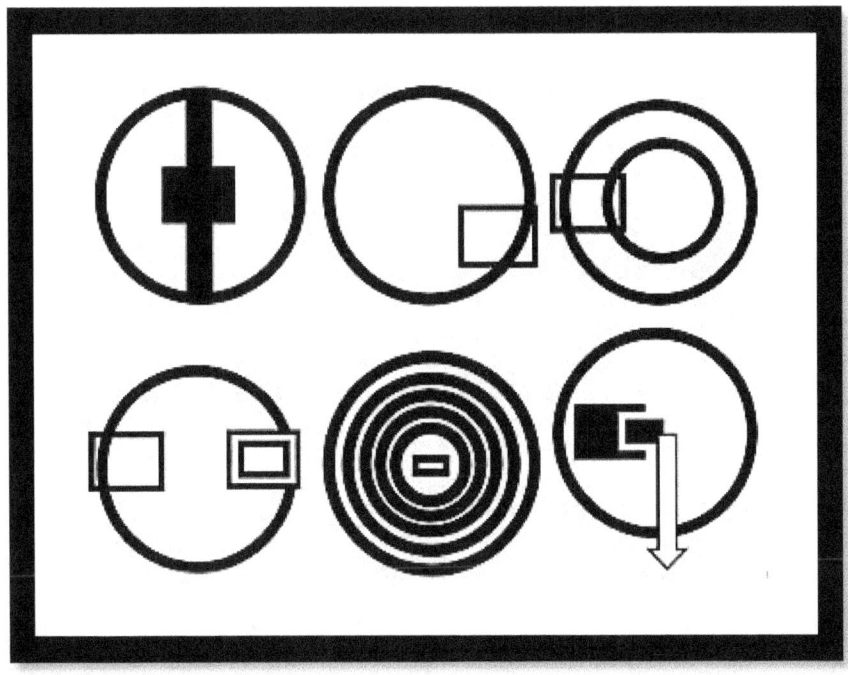

I will ask you to do the following over two minutes – however it will be timed so first listen to (read) the directive.

Starting the first sigil, (GLUE) and working through the final one (BUNDLE AND PUSH) – allow yourself to see them layered upon each other on a clear glass surface. See each one as transparent – only the black parts are dense, the background is clear. Eye movement is helpful. As all of the sigils pile upon one another – you will be able to see through them as if they are lined up to form a telescope. When you see them all – draw the image you see as your composite. There is no RIGHT ANSWER here. Do this for 2 minutes – starting NOW.

(READERS USE A TIMER)

While this two minutes is going on, Sheeg will now walk around and look at you. She will make contact with you. It will be brief, **allow** for that to happen, and allow her to exit in the way that she does. Dissolve the ALL WHITE; take a moment to draw the composite sigil you see, and to write down any Thoughts or questions that you have.

Are there any questions? If there are no questions, then I'm going to say the following things. The process that we just performed, the process of walking through these six sigils in the **ALL WHITE** and allowing for interaction with Sheeg was taking you into a place where you're operating on three levels at the same time. In the way that you look outside of yourself, in the way that you allow for something to look inside of yourself, and in the way that you act in that natural human trepidation of allowing things to insert themselves or come in and out of you. Asking you to Telescope the sigils appears at first to make the situation MORE complex. However it does not. It streamlines and customizes the sigils for you. Now you have a composite that immediately brings you through this ALL WHITE experience – and becomes the TRIGGER for being able to drop immediately into the ALL WHITE when you require a boundary-LESS place to override any doubt you have.

While all that's going on at the same time. The thing that was happening is; those six sigils work in a cooperative way with one another, and allow for the **production of a void in a large ability**. I would like for each of you (said to my students) to tell me what you saw and to send to me the setup of what you experienced with Sheeg, and what you saw with Sheeg, and when you felt she might have been wearing around her neck, and then when I get that information I will be happy tomorrow to tell each of you what Sheeg was wearing around her neck. (readers can write to ask via email).

So now you have a workable immediate customized TRIGGER to a place where you cannot operate in boundaries.....

Thank you, and will meet again tomorrow.

TRANSCRIPT OF SESSION TWO

WNCC – 2-22-17 – In the White Room II

It is the 22nd of February, 2/22/17. I hold very little stock in numbers except for the numbers that reflect what my Daemons are doing, but you really can't ignore 2/22… it feels very solid.

So a few things; first of all, a lot of people are feeling unwell and I'm going to address that, and then I'm going to talk about what unwell means in a larger capacity, and then I'm going to connect that to sexual blockages… and I'm going to put all that into what the ALL WHITE is.

So over a week ago one of you asked me a question about; what is the ALL WHITE? And I said I was going to leave that alone until we met again as a class here.

So the ALL WHITE; first I'm going to say; so the ALL WHITE… and I'm going to leave it there for a minute. And then I'm going to talk about the COLLECTIVE CONSCIOUSNESS of unwell, because when we live around children who go to the heinous Petri dish of school everyday, people pick up all manner of illness via that particular conduit. COLLECTIVE CONSCIOUSNESS interests me because I said something kind of funny to one of you who was unwell. I said perhaps you caught it from XXXXX (another student who was feeling un-well) – and as the classes are on line – there has not been any physical contact among the students. So of course that isn't actually possible, but is it? Is it not? I don't know, but the point is that phrases like "it's going around" are really important to look at. Phrases like "oh everyone has it" are very very important to look at, because the most powerful place we can be in a COLLECTIVE CONSCIOUSNESS is on the productive side of it. If the COLLECTIVE CONSCIOUSNESS is that everyone has a terrible flu, then the right side of it is **outside** of it. If the collective consciousness is that everyone's doing profoundly well in their career, then the right side of the COLLECTIVE

CONSCIOUSNESS is on the **inside** of it.

So I suggest that you put a note to self, suggesting that **everything that appears to be elevated into a state of COLLECTIVE CONSCIOUSNESS affords you an opportunity to choose a side**, because being outside the COLLECTIVE CONSCIOUSNESS when necessary such as; 'everyone has the flu', is a powerful choice. You make that choice based on the fact... and it is a fact... that you've made a decision - so there are many, many layers here... that it is unwell for you to be unwell. It is unnecessary for you to be unwell. It is inappropriate in this time period for you to be unwell.

I had a family member who put on tons of makeup and high heels whenever ill, (as she did that most the time anyway...) . Others thought she was a crazy old lady... I loved her to pieces, and when she was unwell, the heels got higher and more platinum, jewelry went on, and the lips got redder, and it was fantastic to watch. I don't think she had the structural concept of choosing to be **outside** the COLLECTIVE CONSCIOUSNESS....

but she was someone who fought against what was thought to be "normal" or worse, "inevitable". So to counteract what she did not know was the COLLECTIVE CONSCIOUSNESS, she **CHOSE** some sort of **aggressive** statement, usually a visual one, and decided; I'm not going to be unwell; I'm going to get red lipstick and high heels, and I'm going to traipse around in all of my jewelry and consider myself ready to go out and go somewhere to do something. **It was an interesting thing to watch because it was clearly Sorcery even though she had no idea that it was....** You see I didn't invent Sorcery – I observed it and utilized my life to create a process, or perhaps to harness the process into usable sharable form. So please, as you go through life and experiences observe where COLLECTIVE CONSCIOUSNESS is sitting, and what choice that you're going to make in terms of being on the inside or the outside. It is all about selectivity and CHOICE.

And this brings me to the saw blade of the IRON RING... because it helps all people in sorcery to have a visual concept attached to exercise of ritual, or an exercise of action; such as the exercise of pretending that... of watching people pretend that Sorcery doesn't work, and you being on the other side knowing that it does. So the saw blade of the IRON RING is a very valuable tool that I ask you to, at this moment in history, and this moment in time; revisit. One of you mentioned it to me the other night and I said of course she did because I'm about to talk about it... and there's such a connection between minds in this class. So we're going to look at the sigils that we looked at in the ALL WHITE and we're going to address them in a different way tonight, and I'm going to make them work for us in a way where we are imbedding them, as opposed to allowing them to embed in us.

We are embedding them with various things in our life that were not comfortable with and turning them into saw blades, and using them as tools, and keeping them in our toolkit, if you will; air quotes, air quotes... But how do we do that in the ALL WHITE?

What is the ALL WHITE? It's funny because I was going to do a Facebook status about the ALL WHITE and then I realized that it would probably be perceived as ALL WHITE... like ALL WHITE in the cultural way... and I laughed about it and I said; I just am choosing to not be part of **that** type of collective consciousness. So instead there will just be a book because that's the way to do it.

But the ALL WHITE is not actually a place so much as it is a time...an ADDRESS on your timeline. I say address because we cannot easily grasp a way to deliver something, or ourselves, to a time, however we have a complete grasp of delivering something, or ourselves to and address. and this is the meat of what it is that I feel is important to share with you.

If you think about your timeline, and we talk about **addresses on the timeline**, and we talk about placing things in addresses on the timeline, we are not actually placing them on the street that you grew up on. We are not going to place them in a house… we're placing them in a location based on when something happened. If we're going back to 1982 where you decided not to take that girl's phone number and we are going to create an environment where that woman passes through your life again and you get a second shot, we are not going to your timeline to look for where you were standing per se, when that woman tried to hand you her phone number, we're looking at where you were… and that is an issue of time, not location… although we do refer to it as an address in terms of your timeline.

So the ALL WHITE; and we're talking a lot about nuance, and that's why this class is so fantastic and I cherish it so greatly; the ALL WHITE is that Nuance moment/seconds/tiny sliver - whichever language you like about where the arrow hits the target that exists on your timeline when your mind…

....when your mind understands that we are dealing in time and not address. We must grasp that they are essentially the same thing at the foundation level but they are not the same thing when we start to understand how unweighted, how without physical structure, how... I'll use this word *ethereal*, although I hate it... cloudlike if you will, the ALL WHITE is.

> **When you are in the ALL WHITE, unlike being in full usage of the very structured BLACK HOUSES, you personally -and the environment are**
> **100% unstructured.**

When we talk about going into the black houses, I set them up in such a way where walking through a door, you are having an experience, there are things there for you to see and there is a tangibility about it. When we go to the ALL WHITE, everything is possible, and we just place it all out there, and it has to find... **it has to find**... like Sheeg had to find...an address, which isn't an address at all, which is really an **unregulated moment** in time on your individual timelines for you to see her. If I sent Sheeg to your house; if you had a problem and you asked me to send Sheeg to your house, I would send Sheeg to your house and in real time you would see her and something would occur based on our previous conversation.

That's not what happens in ALL WHITE. **In the ALL WHITE she enters and looks at your timeline and sees where she can fit in.** She might've been there for seven minutes before you saw her, she might have just entered and you saw her right away, she might be there the whole time and you only saw her for the middle minute. So what she would do in a circumstance like that is look for the... and I mean this in the best possible use of the word... the **vulnerability**, which is really a **availability**, which is really a place in which you allow for her to show up in your timeline.

So when we are in the ALL WHITE, all of the boundaries, the constrictions, the rules if you will - are gone. That's why we have this very specific set of sigils that operates in the ALL WHITE because they are in a relationship with adherence, and –

> **…..when we are in a relationship with adherence - that's a necessity; to be evolving in sorcery in a place that has no adherences… in other words, you must know what something looks like in order to avoid it…..**

If you do not know what a gun looked like, and a person took out a flask; a shiny flask, I don't know… an oil lamp, a shiny oil lamp, a silver thermos, and an aluminum 45… you would see four things of a similar exterior presence, you would see four things that are of a certain metal structure, **and if you didn't know what they were you wouldn't know which one was dangerous** per se.

That's where you are in the ALL WHITE, but what happens is; you get to redefine your own parameters. You get to say; you know in here I'm not afraid of anything. In here are the ideas of being overtly sexual, for example if that has been something that is difficult for you – in the ALL WHITE it is full on point. This week a lot of you've talked about sexual issues with me; things that are blocking you, and that sort of thing.

> **You can "effigy" yourself in the ALL WHITE and then get to know that person and step in.**

That's one of the things we are going to do, but we're really going to do that one next week. Tonight we're going to work with the sigils again. (NOTE – I am considering a SEDATION on SELF-EFFIGY if there is interest).

But have this concept, have this understanding, have this experience… If you are person who is afraid; and I'm really using that word tonight just as a structural content to communicate sensation, because we all know what fear really is. If you're a person who is afraid/wrought with trepidation - to be overtly sexual for example, and yet has some kind of lower, sort of buried requirement or desire for the right to be, or the right to **experience** being overtly sexual, then in the ALL WHITE you can present the new version of yourself and that new version of yourself will sort of bob and weave with you until it fits into your timeline, and then you'll meet up with it. I suggest utilizing Directional Sorcery and working WEST here so that the work is in the most solid form of APPROPRIATE TIMING. It will be anyway – because the ALL WHITE is set that way – however, the human mind is (and yes I have said this many many times) the human mind is satisfied by equation – so if the equation includes "ALL WHITE **+** WEST=guaranteed timing" – then you are likely to feel doubly insured in the action.

The ALL WHITE is a very complicated place. Usually when I say something like that I say; except but it's not... I'm not saying that this time. In the ALL WHITE there are complications. There are moments where you will see yourself coming and going which is why I don't suggest that you spend a lot of time in it until we have a better grasp of it together or until you feel fully confidant.

In the ALL WHITE mimicry is key. You can create versions of things that are slightly different than the original, (nuance) which is a different kind of mimicry.

For example, if you desire to have published books, you can mimic my experience and have published books. If you desire to have published books that are published by a very large publishing company as opposed to the way that I do it, then you can take my experience and put it in the ALL WHITE, and alter it...

So it's 3-D with a twist. There's really not another place (that **I have experienced**, because there probably is…) I only know what I know…. so that being said –

> **There is not another place that I've experienced that allows for the selective editing of mimicry with this precision.**

> **The ALL WHITE is a wide-open chasm; a void where you decide what's heaviest, what's most familiar, and what's closest, and therefore you control the re-fill.**

So what we are going to do tonight is walk through the sigils and you're going to place your illnesses, dis-eases, dis-comforts, dis-contentions, disappointments, toxins, loneliness, discontent moments etc. in the sigils. Now you all know these sigils well as we used them last week, however I will place them here again for your convenience.

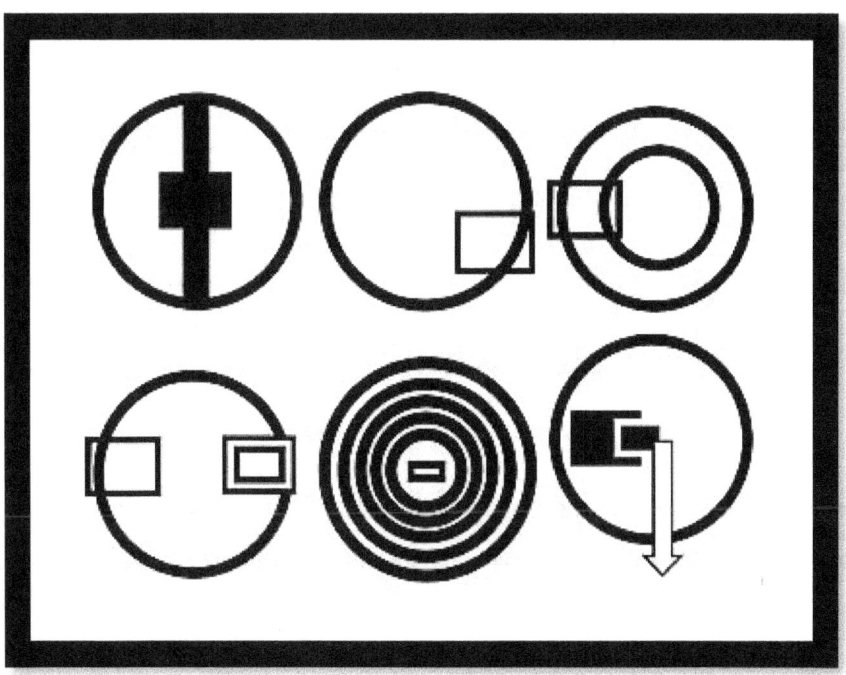

We're going to imbed these sigils with our **GARBAGE**, if you will, and we're going to use the garbage in a specific way, and yes I'm going to explain what I mean by 'garbage'. We're going to use the garbage in a way where we allow for it to be dangerous to others if we choose to use it that way.

> **A note of delicious caution....If you choose to lure someone or something to you for a short or specific period of time, be sure that you have constructed that in a way where there is an out clause or exit strategy.....**
> **That is very important to plan.**

All of you know I do not do "love magick" (note ironic K). The reason – well there are many reasons why I don't do "love magick, but I feel that the **baseline** reason that I don't do "love magick" is because most people cannot construct; I mean, I can... but most people cannot construct the exit strategy. This group can probably construct the exit strategy, however most cannot and are too naive to think it through. If one constructs an artificial emotional attachment – the void around that for the target will constantly re-fill with this artificial emotional attachment created by Sorcery. This will **absolutely** result in obsession. There is that ridiculously "crafted" book, that wonderful "appropriated" story (read mythology and histories of tribal cultures) in which I find so much watered down trivial commentary, called Pet Cemetery. Silly as that is; because I'm a silly girl, and you all know that - Silly as that is, there is a statement in pet cemetery where the author says; and of course, I'm paraphrasing; "Well you can make him come back, but you can't make them come back the way he was... you don't get the same thing." Love magic ends up like that. You ask for something that turns ugly – turns into something desperate. However, in the ALL WHITE you can create interactions where one might almost call it flinging the work. You can create interactions for short experiences you can create interactions that come into place only when you present yourself physically in that experience. THEN THEY CAN EVOLVE in a more experiential and natural way from there....

If you are working in a group of people about whom you desire to create a boundary - about whom you want to create an **ARTIFICIAL FLOAT** – such a way for these people to forget you exist outside the work place – for them to think about you, or even bother you **only** if you are present in the room, you can create that sort of dual level experience (ARTIFICIAL FLOAT) in one sensation of mimicry in the ALL WHITE.

We're going to work with it for a while until everybody gets it down. If you find... and I'm giving you examples right now before we go to the sigils... if you find that you cannot up your game, you cannot think about having more than you need, you cannot think about requiring more than paying the rent, you cannot allow yourself to think of yourself as somebody who looks 25 and is a size 6. If you cannot allow yourself to think about those kinds of things because you desire them and are filled with trepidation regarding the pain of NOT having them – **THEN YOU WILL BE PLACING EFFIGIES IN THE WHITE**; the ALL WHITE will catch up with you, run alongside you, until the right moment comes and they will enter. Again – consider working WEST.

That is some crazy powerful shit... and we're going to do that. We're going to do all of those things. So tonight will going to do is put the sigils up and walk you through a moment for each of them where you allow; **and really, really allow... this is necessary** - Allow the sigils to have a magnetic property if you will, to the junk that you still carry, and when you do; when stuff comes out of you and gets into these sigils, they will remain in the ALL WHITE as IRON RING **saw blades**, which you can use later.

Here is a useful example. If you have a relative in your life who has been particularly torturous to you, and yet seeing that relative, and everyone's had some kind of version of this experience; seeing that relative makes you feel **less** powerful; the way people sometimes are for some reason when they go home and see the parents and suddenly they're 7 or 17 again, … and they give up their power. Or if you are someone who won't bring your lover home to stay in the house because you think they have no adult rights in the presence of family; (all of these kinds of things), then the ALL WHITE is your solution. When you go into the ALL WHITE, you will be **allowed** to do or be whatever you desire because the ALL WHITE allows it, almost chemically so; allows you to stand in front of these sigils, each one, and have things cleared out of you and go into them so that later when you go home and see your parents you can take one of these saw blades with you and cut down all that crap and go be who you are.

Now there are much more important environments than going up to see your parents. I'm using that as an example because it's one everyone can understand. If you feel the following things, tonight is a place to dump them. If you feel unwell, if you feel oppressed, if you feel sexually oppressed, if you feel that you are under the control of a circumstance that you cannot even **begin** to unpack because it seems so complicated, if you feel that you are in service to something that you no longer desire to be in service to; those are the Ideas (and they are Ideas) that I'd like you to think about as we go through a few minutes of timing for each of these sigils, and allow them to fill themselves with whatever they pull out of you.

We are going to go in, we are going to allow each sigil to pull out, and we're going to be in the ALL WHITE together. I'm going to place my Attending Daemon 8 in the ALL WHITE. 8 is sort of a miasma of fog which will make the ALL WHITE even odder to you and perhaps make it feel like the space is filled with more hydrogen than one might need, and in that experience they'll be a sensation of almost being a little high. This process will allow the stuff to get out of you, because it has to get out of you, and stick to the sigils, but this time you will own the sigils; they will still be there for you. And I suggest after a sigil has pulled everything out, that you take a few notes on what it is that you saw go into the sigils. Then burn the notes. Fuck it. You don't need that crap.

I ask you not to preconceive this. I ask you to get in that nuanced place where you **have to accept** that there is an opportunity for this to work for you, not because of you. I set it up that way, so that you are not burdened by having to make these decisions, but rather, you are going into the diagnostic situation and the machine is pulling out the garbage. So that being said I'm going into my ALL WHITE place this so you can see it. Then I will time each of these sigil appearances for 2 ½ minutes followed by 30 seconds of writing. (READERS OF THIS – SET PHONE TIMERS FOR 2 ½ minutes and follow that with a short bit of writing).

Now 2 ½ minutes is a long time and it might feel long, and it might feel uncomfortable, and that would be **wonderful** if it did feel **uncomfortable**, because what we're looking for here is the sucking machine; for it to pull out of you everything that it can so that two things happen; you are without the burden of those things, and therefore lighter and able to travel per se, and that we are creating virulent saw blades out of these sigils.

Here is another image of the sigils for your convenience.

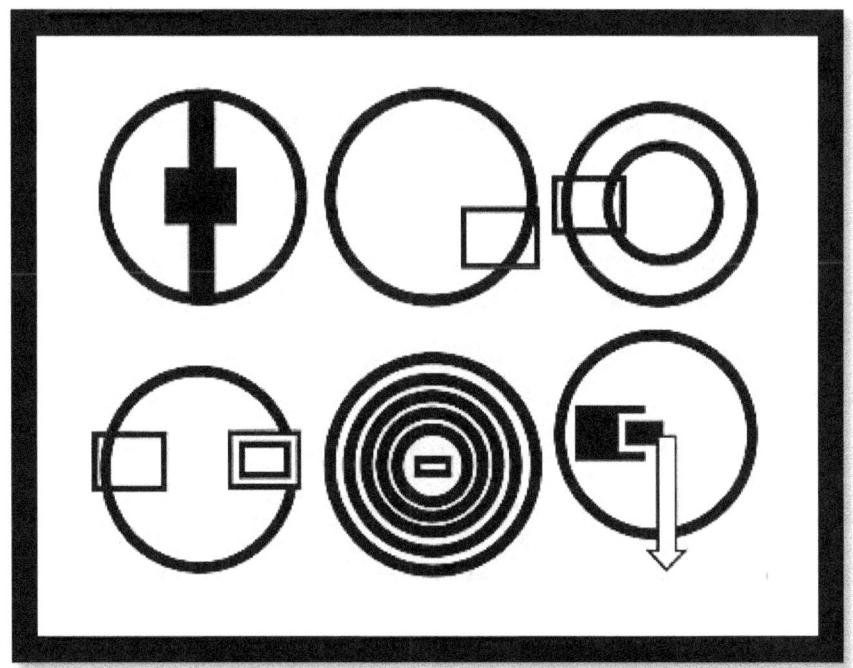

We will start with **GLUE**, and I will time you for two minutes, and then a half minute of writing. Okay so here's what you're going to see; you going to go into the ALL WHITE and in the ALL WHITE I'm placing 8 in the ALL WHITE. 8 has a sensation of being very much like a fog, or mist; a miasma, but with a density that is showing you that there is clearly a flexible entity in the space. Now remember, you are in the ALL WHITE; 8 will catch up with your timeline as is required. Ok, everybody seems to be in it, and feeling it. If your arms get itchy it's okay, 8 sometimes causes that. you might feel a sensation of dampness around him.

The ALL WHITE does not have walls, it is not a room, it is an address in time.

It is an address in time. I'm teaching this class to operate in time. And speaking of time, I will start the timer now with the first sigil which we refer to as **GLUE. (READERS START YOUR TIMERS)**

I'm seeing that some of you are having a sensation that you don't want to get rid of something - that you don't know why you don't want, get rid of it- so I suggest you let go no crutches. Stop and write a few notes please. I will pause the screen, and I'd like you to stay in the ALL WHITE and allow the sigil that you just used to saw blade. Allow interactions with 8. Stop write anything else you feel you must write in just the next few seconds.

And now we begin the process with the second sigil, PARANOIA.

(READERS START YOUR TIMERS)

Begin. yes you are actually seeing the things that you think. the same stop and take a few notes. And now for 30 seconds well this set this sigil to saw blade, and to remain your space. Utilize 8. (Timed - stop.)

Before we do **MASQUERADE** I'm sending to you a Disincarnate performer. That's all I'm going to say, I will tell you it is a woman and that she is known to have been quite elaborate in her presentation. And she enters this process to bring irreverence, beauty, decadence, sexual energy, and overt physicality. 8 is still in the room, in this space; in the ALL WHITE. Allow, allow, allow, and begin the work with MASQUERADE. **(READERS START YOUR TIMERS)** In the presence of MASQUERADE sometimes time it feels longer, this one might feel longer. Use the format of the left box. Is the box escaping or is the circle? 9 himself will walk through the ALL WHITE as he seems to be required to do so at the moment. We are observing the placement of your saw blades; making sure they remain steadily in the ALL WHITE. Allow MASQUERADE to spin and be a saw blade. You might notice that MASQUERADE likes to store itself in odd locations; sometimes behind you, sometimes under you. Times up – now go ahead and write anything you need to write.

Now we begin the process with **COMPARISON**.
(READERS START YOUR TIMERS) Allow-Allow. COMPARISON is the one where we become trapped in the methodology of comparing ourselves to others of progress to others all this sort of thing. That is sometimes useful and sometimes not, the point is to be in control of the technique, not to judge the individual circumstances utilize COMPARISON - allow it to take from you everything it can possibly relieve you of. You will most likely find that time is quicker in COMPARISON. The Disincarnate has stepped away. Time is up. Write some notes, and allow for the sigil to act as a saw blade for 30 seconds.

We are beginning now with WONDER. First I will say that I'm inviting into the space a Disincarnate that is thought to be a great healer. He is somebody I actually knew personally, who is now Disincarnate. he is in the space, he has a childlike wonder about him and is an older man. Listen for his voice if he desires to tell you anything**. (READERS START YOUR TIMERS)** Start.

You'll find that time moves slower in WONDER. Time is up -Stop. Also allow WONDER to saw blade. Stop. Note the location of the saw blade. Write down whatever it is that you'd like to write down.

We have one more to do, this is **BUNDLE AND PUSH.** Just use it the same way, we're not using it any differently than the other ones. Just allow whatever is left to go into bundle and push; these things are in order that they are in for reason; bundle and push having the strongest magnetic content... it will have a stronger pull. Whatever is left is going into bundle and push. **(READERS START YOUR TIMERS)** Start now. If you are unwell really let it go. Utilize the **path** in **BUNDLE AND PUSH**, utilize the area; utilize the open white area so that things can be pulled out and drawn down in a way. Time moves very slowly in **BUNDLE AND PUSH**. Stop. The Disincarnate that I sent has left your area. Allow for **BUNDLE AND PUSH** to saw blade. Write down anything else that you desire to write.   Then take a moment or so to come back to this place.

I'm going to take this opportunity to talk for a few minutes about the usefulness of this as we have come away from the ALL WHITE. As you leave the ALL WHITE, the question is do you leave it? As the presence of 8 dissipates, which I am instructing him to do now, and you step away from the ALL WHITE, do you leave it - or do you release it for a period of time? Please note (recall specifics or write) your experience of leaving the ALL WHITE, or releasing it, and note that when we revisit this next week, we're going to begin with **effigies of ourselves**. (NOTE – there will soon be a Sedation on this) **However, I mentioned that the ALL WHITE is about time. It will catch up to you on your timeline as you require it.** Even the saw blades themselves will evolve with you. The things that are connected to the things that you let go of tonight will grow into those saw blades. **As we proceed in the ALL WHITE you will see that it moves along the timeline with you, so in a sense the ALL WHITE is where you are, and yet allows you to reach back in and forward in some cases to make alterations.** 8 has pulled out, and I've pulled out my participation in your ALL WHITE.

Q&A

Does anybody have any questions?

STUDENT - should be going into the ALL WHITE to do all our installations in all of our Sorcery now?

S9 – No... it's an option... we do major modifications in the ALL WHITE. These are modifications. We create **toolkits** in the ALL WHITE... for example - the saw blades are toolkits. You will start to feel where the ALL WHITE fits in, the same way you learn to feel where the **STATIC PRACTICE** does. It will have better applications than other structures because it is fluid and is a perfect place for SELF-EFFIGY and mimicry. We are going to do a lot of work in the ALL WHITE so that everything we create remains... Everything is 3-D printed there. But you will not need to enter it every time you do sorcery.

There will be a lot of things that come out of it over these next couple of weeks that allow you to have a permanent toolkit of some of the things that you will be using as you get into more advanced work. Does that answer your question at all?

STUDENT - yes it does. Can you compare and contrast this to the static practice?

S9 – well I can - however... we're going to get there, and the reason I am sort of spacing it out is because the ALL WHITE takes time to settle. It is actually an entity in a way. So ask me that question again... perhaps the next time. However I will say this – and this is a good beginning to the answer to your question - the STATIC PRACTICE is made from your DNA and your life experience in its totality. The ALL WHITE is made from your experience as a baseline – and your desire to ALTER your experience through mimicry, SELF-EFFIGY and other techniques. It is built, the STATIC PRACTICE simply IS. Does that make sense? And in a further contrast, The **BLACKHOUSES** are sort of the other dwelling on your property where you go to do things that you don't want to do in your own house because there are intentional toxins in that work as you all know.

The **ALL WHITE** is a place that's built entirely of your experiences. The **STATIC PRACTICE** is all built from your DNA. When you have all three of these locations in place, there is very little that I can imagine that you can't do. Part of what Sorcery is, is knowing where to perform it, and this will give you a level of options that you didn't have before.

If no one else has questions I will go for the evening. You will feel a certain amount of… I would say moments without crutches; the confusion of not having the ugly controlling parts of your experiences to rely on. By "rely on" I mean they are no longer an out – an exit strategy for not completing whatever brings you success and Pleasure. **You will notice that you will start to rely more on the moment that you're in as opposed to connecting to all the terrible things that have happened to you in the past that reflect how you would usually see that moment.** It is not a new filter – it is no filter. All fresh views…Allow that to happen, write about it, write emails to me, and we will pick this up next week.

IF YOU ARE INTERESTED IN THE POTENTIAL SEDATION ON the enormous use of SELF-EFFIGY, CONTACT THE SORCERESS AT sorceresscagliastro@gmail.com

CURRENT STUDENTS - PLEASE USE THE STUDENT EMAIL ADDRESS

If there is enough interest....the SEDATION will be scheduled........S9

To contact the author, send enquiries by email to

northseatales44@gmail.com

SORCERESS CAGLIASTRO,
in the hands of 9

Author, Publisher, Teacher, Pleasure Enthusiast, Blood Sorceress, Necromancer, Creator of the Iron Ring Science of Sorcery and the foremost authority on Blood Sorcery, Parent, Explorer, owner of THE FIRM which is a group of elite Sorcerers for hire, Publisher of RESCUING KNOWLEDGE PROJECT (search RESCUING KNOWLEDGE PROJECT on Amazon for newest titles) which reprints esteemed literary works long forgotten for new audiences......

SORCERESS CAGLIASTRO, Blood Sorceress, Necromancer... gratefully in the hands of 9

www.cagliastrotheironring.com

Blank for notes...

Blank for notes...

www.ingramcontent.com/pod-product-compliance
Lightning Source LLC
Chambersburg PA
CBHW070347230526
45471CB00006B/2452